CAMBRIDGE
POCKET
STAR ATLAS

JOHN COX

W9-CTI-407

Contents

CAMBRIDGE
UNIVERSITY PRESS

The Author would like to thank Mark Jenvey and Caroline Rayner for their help in preparing the revisions to this Atlas.

Published by the Press Syndicate of the University of Cambridge
The Pitt Building, Trumpington Street, Cambridge CB2 1RP
40 West 20th Street, New York, NY 10011-4211, USA
10 Stamford Road, Oakleigh, Melbourne 3166, Australia

First published in Great Britain in 1993 by George Philip Limited
Second edition published 1996

First published by Cambridge University Press 1996

Library of Congress cataloguing in publication data available

This edition only for sale in the United States of America
and Canada

ISBN 0 521 58992 4 paperback

Printed in Hong Kong

Introduction

The *Cambridge Pocket Star Atlas* is intended to help a new observer to identify the stars, planets, and constellations, and to serve as a minimum atlas for the traveler and the more experienced observer.

The best conditions for observing the stars and planets are dark, clear skies, when the stars appear brighter and more stars can be seen. A bright moon washes out the fainter stars. The street lighting used in towns and cities creates particular difficulties for the urban observer, but the planets and the brighter stars of the major constellations can be seen in the middle of a city provided the observer chooses a clear and otherwise dark night and avoids having nearby lights shining into the eyes: in other words, observes from somewhere in local shadow. Low-cost telescopes advertised as "suitable for astronomy" are often of little practical use, and are frequently a source of disappointment. Low-magnification binoculars (7x or 8x) are by far the most useful kind of telescope for the beginner, and have many specialized uses for the more experienced observer.

The celestial sphere

All the objects of the night sky – the stars, the planets, and the Moon – are at enormous and varying distances, with the more distant objects thousands of millions of times more distant than the nearer, but to the human eye they all seem to be equally far away, as if placed on the inside of a gigantic "celestial sphere" that encloses the Earth in the way that a shell encloses an egg. This celestial sphere is an illusion, but it is also a useful convention, and it is used to describe where celestial objects appear to be.

Over the course of the night the stars appear to rise in the east and set in the west. This is an apparent motion produced by the rotation of the Earth; a more familiar manifestation is the apparent motion of the Sun over the course of the day. At night the Earth's rotation makes it look as if the whole sky is moving as a piece, as if the imaginary celestial sphere is rotating round the Earth.

Celestial coordinates

The position of a celestial object is given by a system of "celestial coordinates" that work like terrestrial latitude and longitude projected on to the inside of the (imaginary) celestial sphere. The points overhead of the terrestrial poles are called the north and south "celestial poles." The circle of sky that passes overhead when observed from the terrestrial equator is called the celestial equator.

The angular distance of an object north or south from the celestial equator is reckoned in degrees of declination. Its angular distance around the sphere is measured in hours and minutes of right ascension (RA). Right Ascension is reckoned eastward of the position of the Sun as it crosses the celestial equator at the time of the equinox in March: the vernal equinox for the northern hemisphere of the Earth. This position is called the "First Point of Aries." There are 24 hours of RA altogether, so each hour of RA is equivalent to 15° of arc.

The fixed stars

The Earth is a member of the solar system, one of nine major planets in orbit round a central star, the Sun. The Earth orbits the Sun at an average distance of 150 million km (93 million miles). Light, travelling at approximately 300,000 km per second (186,000 miles per second), takes eight minutes to travel from the Sun to the Earth. The nearest naked-eye star beyond the Sun is Alpha Centauri, lying at a distance of 4.3 light-years, which is to say that light takes 4.3 years to travel from it to us. Most of the stars visible to the naked eye (meaning visible without telescopes or optical aids) are very much farther away.

When the positions of stars are carefully measured through telescopes over a period of years, it can be worked out that they are moving about in independent directions at high speeds. But their distances from us are so great that to the naked eye they appear to be "fixed" in the same positions relative to one another. Change will eventually become apparent, but over a few thousand years the relative position of the stars looks the same.

4

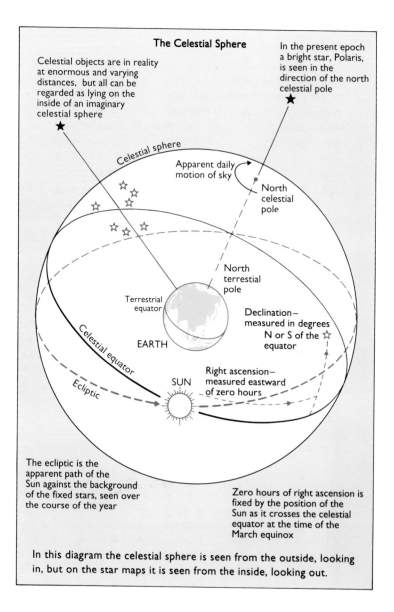

The Celestial Sphere

Celestial objects are in reality at enormous and varying distances, but all can be regarded as lying on the inside of an imaginary celestial sphere

In the present epoch a bright star, Polaris, is seen in the direction of the north celestial pole

Celestial sphere

Apparent daily motion of sky

North celestial pole

North terrestial pole

Terrestrial equator

EARTH

Declination— measured in degrees N or S of the equator

Celestial equator

Right ascension— measured eastward of zero hours

Ecliptic

SUN

The ecliptic is the apparent path of the Sun against the background of the fixed stars, seen over the course of the year

Zero hours of right ascension is fixed by the position of the Sun as it crosses the celestial equator at the time of the March equinox

In this diagram the celestial sphere is seen from the outside, looking in, but on the star maps it is seen from the inside, looking out.

The ecliptic

The Earth takes 365¼ days to orbit the Sun. From the viewpoint of the Earth it looks as if the Sun is moving along a path traced out against the background of the fixed stars – or at least, it would if the stars could be seen while the Sun was in the sky. In fact the position of the Sun against the stars can be directly observed only during a total solar eclipse, when the brighter stars become visible. The apparent path of the Sun is called the ecliptic, taking its name from being the path on which eclipses of the Sun and Moon take place.

The Moon and the planets all appear to circle the sky within a few degrees of the ecliptic, and their positions along it may be measured in degrees of "celestial longitude," reckoned eastward from the First Point of Aries (0°). The Sun appears to move 1° eastward along the ecliptic in 24 hours, an effect of the Earth's westward motion in its orbit round the Sun.

The axis of the Earth's spin is not square to the plane of the Earth's orbit round the Sun – the plane of the ecliptic – but is tilted over at an angle of 23.45°. The plane of the Earth's axial rotation is regarded as primary, so the angle between the planes is known as the "inclination of the ecliptic" and is instrumental in a number of effects, including the seasons of winter and summer in the temperate regions of the Earth. For half of the year one hemisphere is tilted toward the Sun, making the long days and short nights of summer; for the other half of the year the same hemisphere is tilted away, making the short days and long nights of winter.

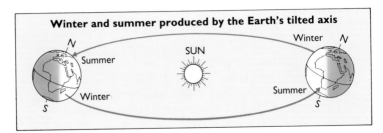

Winter and summer produced by the Earth's tilted axis

In summer in the temperate regions of the northern hemisphere the Sun rises above the horizon at a point north of east, passes high through the sky, and sets at a point north of west. In winter the Sun rises south of east, passes low through the sky, and sets south of west. In the southern hemisphere the pattern is reversed, with the Sun rising north of east in winter, and south of east in summer. The most northerly and most southerly rising and setting positions, and the longest and shortest days, are reached at the times of the solstices (around June 21st and December 21st).

At the time of the equinoxes (around March 21st and September 23rd) the Sun passes directly overhead of the observer at the terrestrial equator. Observed from anywhere on the inhabited earth it appears to rise due east and set due west, and the lengths of day and night are exactly 12 hours each (hence the name "equinox," "equal night").

Meridian passage

A "meridian" is any great circle that passes through both of the poles. The phrase "observer's celestial meridian" describes an imaginary line that starts from the horizon at due north, passes directly over the observer, and meets the horizon at due south. The Sun crosses this meridian at true local midday (from the Latin *meridies*, "midday"). When a celestial object crosses the meridian it is said to be at meridian passage. Other terms which describe the same event are "culmination" and "upper transit."

Circumpolar stars

Except for those living close to the terrestrial equator, observers will have a view of either the north or the south celestial polar regions. Stars close to the poles appear to move round in a circle over 24 hours, and never rise or set. These are called circumpolar stars. A star crossing the meridian above the pole is said to be at upper transit, while the same star crossing the meridian below the pole (12 hours later) is said to be at lower transit.

Time

Mean (average) time – ordinary clock time – is based on the average length of a day and a night. It takes 24 hours on average to get from one midday – one meridian passage of the Sun – to the next. In the same period the Earth travels nearly 1° in its orbit round the Sun; the Sun appears to move 1° eastward along the ecliptic, and in order to bring the Sun back to meridian passage the Earth has to rotate through 361°.

"Sidereal time" is defined by the period that is required to bring the same star from one meridian passage to the next. To bring this about the Earth has only to rotate through 360°. This defines 24 hours of sidereal time ("star time," from the Latin *sideralis*, "of a star"). Twenty-four hours of sidereal time takes up only 23 hours and 56 minutes of average mean time. Measured by the clock, the same star returns to meridian passage four minutes earlier every night.

From one night to the next the same stars appear higher in the eastern sky, reach meridian passage sooner, and set sooner. The observer looking out on the sky at the same time every night will notice that over the course of the year the area of sky that is open to view moves slowly eastward over the course of the year. The map on pages 24–25 shows when particular areas of the sky cross the meridian.

The constellations

From the earliest times observers have in their imagination joined up apparent groupings of stars into the outlines of giant figures, the constellations. In reality most of these figures include stars at enormously different distances, with no true association in three-dimensional space.

The major constellations of the northern and equatorial sky are a combination of Mesopotamian and Greek figures, listed by the Greek astronomer Hipparchus in about 130 BC, and transmitted in the *Almagest* (*c*. AD 140) by the Egyptian astronomer and geographer Ptolemy.

Two thousand years ago the ecliptic was divided into 12 sectors, each 30° wide, and named after the constellation found in it. The Sun entered the First Point of Aries (the beginning of the sector containing Aries) at the time of the vernal equinox. One effect of the Moon's gravity is to make the axis of the spinning Earth gyrate slowly, completing one gyration in 25,760 years. The gyration produces no change in the plane of the ecliptic, but it creates a slow change in the Earth's orientation toward the stars, thus changing the plane of the celestial equator, and causing the First Point of Aries to move slowly westward along the ecliptic. Because of this precession each sector of the ecliptic is now almost 30° removed from its original position.

Minor and southern constellations were added by Europeans in the 16th, 17th, and 18th centuries. Constellation boundaries were formalized in 1930. Constellation figures were shown by simplified ball-and-link "asterisms" in China $c.$ 200 BC or earlier.

Star names and designation systems

Star names are mostly Arabic, with their English forms changed in transmission and use. The brighter stars in each constellation were assigned a Greek letter by Johann Bayer in the 17th century and a number by John Flamsteed in the 18th.

An extension to the Bayer system assigned lower-case letters, a to x, and capital letters, A to Q, to some stars (mostly found in the southern sky). In the 19th century Friedrich Argelander listed variable stars in particular constellations by capital letters from R onward, in a system that was later extended to double letters and combinations of letters and numbers.

Apart from the stars, which appear pointlike in even the most powerful telescopes, there are a number of more indistinct objects, originally called nebulae (see page 11). Many of the brighter nebulae were listed by Charles Messier in the 18th century, and are still known by their "Messier Numbers," M1 to M109. A subsequent and more extensive listing was made in Dreyer's *New General Catalogue* of 1888, from which "NGC numbers" are drawn.

The Milky Way

The Milky Way is observed as a band of light that crosses the sky, and may be seen over the course of any dark and clear night. Observed through a telescope, the Milky Way is resolved into a swathe of faint stars. Dark patches in the Milky Way are regions of obscuring gas and dust.

The Milky Way is a galaxy system of approximately 200 thousand million stars, together with vast amounts of gas and dust, arranged in a disk-like shape about 120,000 light-years in diameter. The Sun lies in the plane of the disk about 30,000 light-years from the center, and the Milky Way is the disk seen edge on. The plane of the Milky Way is represented on maps as the "galactic equator." The center of the Galaxy (0° of galactic longitude) lies beyond the stars that make up Sagittarius. The nearside outer edge (at 180° of galactic longitude) lies beyond the stars of Auriga. The North and South Galactic Poles (NGP and SGP) are located at 90° to the galactic plane.

The Milky Way is one of an estimated 100 thousand million galaxies in the visible universe. Only four "external galaxies" (galaxies other than our own) are visible to the naked eye. The Large and Small Magellanic Clouds are dwarf irregular galaxies (that is, having no definite shape) at distances of about 160,000 and 200,000 light-years, visible as cloud-like objects in the southern sky. The Andromeda (M31) and Triangulum (M33) galaxies are visible as smudgy objects in the northern sky; both are spiral galaxies similar to our own Milky Way Galaxy, but lying at a distance of more than 2 million light-years.

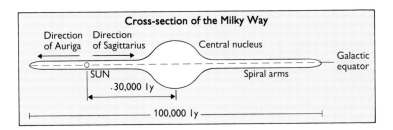

Cross-section of the Milky Way

Galaxies, star clusters and nebulae

The term "non-stellar object" has been used to describe any celestial object that does not have the pointlike appearance of a star: galaxies, star clusters, and gas clouds. Most of the non-stellar objects shown on the maps are rather dimmer than the faintest stars that are shown, but several are visible with the naked eye under favorable conditions, and most make interesting objects for observation through a pair of binoculars.

External Galaxies have been described above. Other kinds of object lie within the Milky Way Galaxy itself. Nebulae describe vast clouds of dust and gas (mainly hydrogen), often associated with areas in which stars are still being formed. Open clusters are groups of young stars that are presumed to have formed together in the same gas cloud and have not yet drifted apart. Globular clusters are densely populated, compact spherical groups of old stars.

Magnitude

The brightness of stars (and of all celestial objects) is measured in a system of magnitudes, a system inherited from the Greeks, who believed that some stars appeared brighter than others because of their size. It is now clear that the intrinsic brightness of stars varies a great deal, and that another important factor that affects the apparent brightness of a star is its distance. "Magnitude" is still used to describe brightness, but it is used in two ways. Apparent visual magnitude (m_V) describes how bright a star appears in the night sky. Absolute visual magnitude (M_V) describes how bright a star would appear from a standard distance (≈ 32.26 light-years), and may be used as a measure of how bright the star is in itself.

In the Greek scale the brightest objects were stars "of the first magnitude"; less bright stars were stars "of the second magnitude," down to the dimmest stars that could be seen, stars "of the sixth magnitude." The modern scale is much in agreement with the original, but it has been extended through zero and into negative figures in order to accommodate the brightest objects, and recalibrated to make a difference of one magnitude represent a difference of just over 2.5 times the brightness (allowing a difference of five magnitudes to exactly equal a difference of 100 times the brightness). Spica, in Virgo, has an apparent magnitude of $+1$. Vega, in Lyra, is 2.5 times brighter at m_V 0.0. Sirius, in Canis Major, the brightest star in the sky, has an apparent magnitude of -1.46, 9.6 times as bright as Spica.

Under optimum conditions the faintest stars that can be seen with the naked eye have an m_V of about $+6.5$. There are about 9000 objects within this limit (shown in a standard reference, *Norton's Star Atlas* Longman, UK; Wiley, USA). Apart from a few fainter stars shown for location purposes, the *Cambridge Pocket Star Atlas* is limited to m_V 4.75, showing approximately 1000 stars, enough to identify the main stars in each constellation.

Spectral types

When the light from a star is passed through a prism (or across a diffraction grating) and focused, its spectrum shows "absorption lines" that correspond to ionized elements in the atmosphere of the star. Analysis of their spectra allows most stars to be placed on what is in effect a temperature sequence that runs from the hottest "O-type" stars through to the coolest "M-type" stars. The complete sequence is O, B, A, F, G, K, M. Each type has subgroups, so that F runs into G through the subgroups F8, F9, G0, G1, G2, G3.... The Sun is of spectral type G2.

When the absolute magnitudes of stars are plotted against their spectral types in the (what is known as) Hertzsprung–Russell Diagram, it is seen that examples fall into a number of discrete groups. Most line up in a "main sequence" of compact dwarf stars (luminosity class V).

A number of stars of exceptional intrinsic brightness are found higher up the diagram. These are believed to be high-mass stars with complicated nuclear reactions that have caused the star to become enormously distended. They are classified as various kinds of giant and supergiant stars (classes I–IV).

Hertzsprung–Russell Diagram

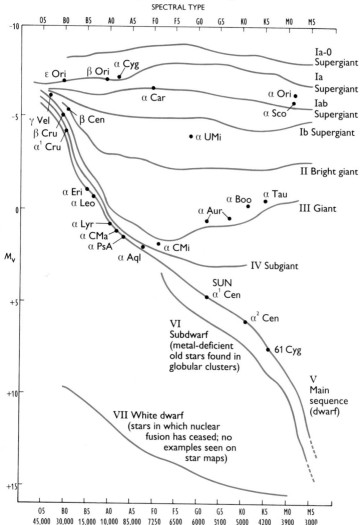

EFFECTIVE SURFACE TEMPERATURE (K) IN MAIN-SEQUENCE STARS

The MK system of classifying stars

The MK system uses a capital letter and Arabic numeral to describe a star's spectral type, Roman numeral I to VII to describe its luminosity class (note subtypes Ia-0, Ia, Iab, Ib) and following lower-case letters to describe any spectral particularities. Luminosity classes I–IV contain various giant and supergiant stars; classes V–VII comprise dwarf and white dwarf stars. Unusual spectral types that are referred to in the tables are WR (*Wolf-Rayet*, extremely hot stars whose spectra show many emission lines), and S (spectra marked by zirconium lines)

Peculiarity codes used in the tables are these:
e: emission lines – added radiation at specific wavelengths
m: metal absorption lines
n: diffuse lines – produced by rapid rotation
p: peculiarity in spectrum
eq: emission with short wavelength absorption
s: sharp lines
v: variable lines

Colors of the stars

The human eye is not sensitive to color in very dim objects (hence the black and white quality of vision at night) but the brightest stars do appear colored to the sensitive observer. Red giants such as Betelgeuse, Aldebaran and Antares are distinctly orange–red in appearance, while hot B and A types like Regulus, Sirius, Spica, and Vega appear blue to blue–green.

Multiple and variable stars

Most stars in the solar neighborhood are multiples, meaning that they are members of a gravity system involving two or more stars in mutual orbit. Variable stars are stars whose brightness changes. Many stars are variable to some extent, but in most cases the range of variability is not very great.

The Moon

It takes an average of 27.32 days for the Moon to complete (what may conveniently be regarded as) an orbit around the Earth, measured in relation to the stars. This is its sidereal period. This period is not the same as the time from new moon to new moon. The phases of the Moon depend on the relative positions of the Sun, Moon and Earth, and it takes 29.53 days for the same geometry to return: this is known as the "synodic period" (from the Greek *synodos*, "coming together").

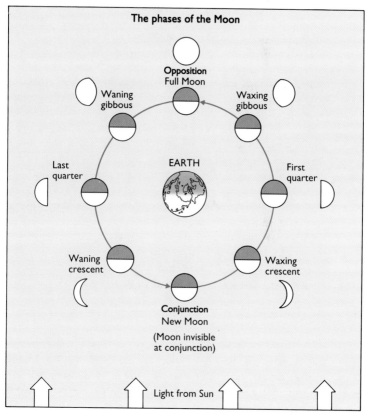

The phases of the Moon

The Moon's rotation on its own axis is "captured" by the period of its orbit round the Earth, so that it rotates once in each orbit and always presents the same face to the Earth. Astronomical new moon is the name given to the Moon at conjunction, when it lies in the same direction as the Sun. At astronomical new moon the side of the Moon that faces the Earth is in shadow and cannot be seen.

The Moon is first seen (as what is ordinarily called the new Moon) on the second evening after conjunction, when it appears low in the western sky after sunset. Observed from one evening to the next the Moon moves 13° eastward against the stars, appearing higher in the sky each evening and setting 50 minutes later. From one evening to the next it "waxes," as an increasing proportion of its face is seen illuminated.

Seven to eight days after conjunction the Moon is seen at first quarter; the name may refer to the quarter-sphere that is seen or the quarter-circle angle (90°) that it makes with the Sun. At first quarter the Moon is at the meridian at sunset and sets at midnight.

Between first quarter and full moon, the Moon appears gibbous (from the Latin for "humped"). It appears in the eastern sky at sunset, and sets before dawn. Fifteen days after conjunction the whole visible face is illuminated – the "full moon." The full moon is at opposition – that is, opposite to the Sun in the sky; it lies at an angular distance of 180° away, rises at sunset, passes the meridian at midnight, and sets at dawn.

After full moon the Moon begins to wane. At last quarter it rises at midnight and reaches the meridian at dawn. The last visible Moon of the cycle is seen usually two days before conjunction, rising before the Sun.

Movements of the Sun and Moon compared

The Moon shows the same patterns of rising and setting that are seen in the Sun, and the same pattern of passing high or low across the sky, but while the Sun takes a year to exhibit the whole pattern, the Moon shows the full range of movements in the course of a (sidereal) month.

The Moon's orbit round the Earth is inclined to the plane of the ecliptic at an angle of 5.15°, and the points of intersection (the nodes) circle the ecliptic westward in 18.6 years. For nine years the inclination of the Moon's orbit is additive to that of the ecliptic, so that the monthly extremes of the Moon's position are greater than the yearly extremes of the Sun. For the other nine years inclination is subtractive, so that the Moon's extreme positions are less than the yearly extremes of the Sun.

Eclipses

An eclipse of the Sun takes place when the Moon passes directly between the Earth and the Sun. An eclipse of the Moon occurs when the Moon passes through the shadow of the Earth. Solar eclipses therefore take place only at the time of new Moon, and lunar eclipses only at the time of full moon.

Because the plane of the Moon's orbit is inclined to the plane of the ecliptic, the Moon at opposition and at conjunction is usually above or below the ecliptic plane. Eclipses can occur

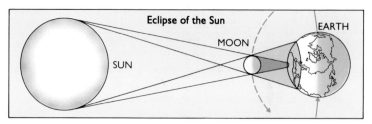

only when the Sun is passing a node at the time of opposition or conjunction. This happens every 173.3 days, and there is usually at least one solar and one lunar eclipse near this time.

The Moon's diameter is hundreds of times smaller than the Sun's, so the Moon throws a cone-shaped shadow. Eclipses of the Sun are more common than eclipses of the Moon because the Earth makes a large target, but the tip of the Moon's shadow tracks over a very small area of the Earth, and total eclipses of the Sun are very rare from any particular place on the Earth's surface. Annular eclipses occur when the distance from the Earth to the Moon is close to maximum (the Moon at apogee) so that the dark moon is seen surrounded by a disk of sunlight. Partial eclipses, in which only part of the Sun's disk is covered, can be seen from a much larger area of the Earth's surface and are fairly common.

Looking at the Sun, even through filters, can cause blindness. Eclipses of the Sun can be observed by projecting an image through a pinhole.

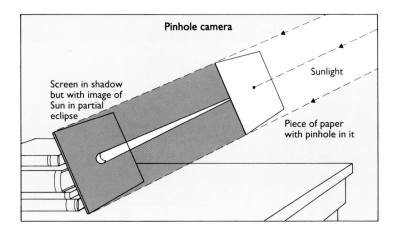

Pinhole camera

Sunlight

Screen in shadow but with image of Sun in partial eclipse

Piece of paper with pinhole in it

Lunar eclipses are seen when the Moon passes wholly (total eclipse) or partly (partial eclipse) through the Earth's shadow, and are visible from anywhere that has sight of the Moon at the time of the eclipse.

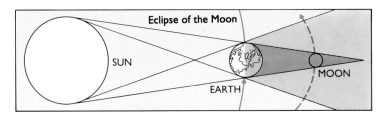

Eclipse of the Moon

SUN

EARTH

MOON

Eclipses of the Sun and Moon 1996–2015

SUN: ●T: Total Eclipse   ●P: Partial Eclipse   ●A: Annular Eclipse
MOON: ☽T: Total Eclipse   ☽P: Partial Eclipse

Date and (approximate) time given as: hour (GMT) / day / month.
Rising node (☊) position at 0h January 1st given for each year,
expressed in degrees of the ecliptic.

1996 ☊ 202.4°	**2001** ☊ 105.7°	**2006** ☊ 9.0°	**2011** ☊ 272.3°
☽T 0010/04/04	☽T 2020/09/01	●T 1010/29/03	●P 0850/04/01
●P 2240/17/04	●T 1205/21/06	☽P 1850/07/09	●P 2110/01/06
☽T 0255/27/09	☽P 1455/05/07	●A 1140/22/09	☽T 2010/15/06
●P 1400/12/10	●A 2050/14/12		●P 0840/01/07
		2007 ☊ 349.7°	●P 0620/25/11
1997 ☊ 183.0°	**2002** ☊ 86.4°	☽T 2320/03/03	☽T 1430/10/12
●T 0125/09/03	●A 2345/10/06	●P 0230/19/03	
☽P 0440/24/03	●T 0730/04/12	☽T 1040/28/08	**2012** ☊ 253.0°
●P 2400/01/09		●P 1230/11/09	●A 2350/20/05
☽T 1845/16/09	**2003** ☊ 67.3°		☽P 1100/04/06
	☽T 0340/16/05	**2008** ☊ 330.4°	●T 2210/13/11
1998 ☊ 163.7°	●A 0410/31/05	●A 0355/07/02	
●T 1730/26/02	☽T 0120/09/11	☽T 0325/21/02	**2013** ☊ 233.6°
●A 0205/22/08	●T 2250/23/11	●T 1020/01/08	☽P 2010/25/04
		☽P 2110/16/08	●A 0030/10/05
1999 ☊ 144.4°	**2004** ☊ 47.7°		●T 1250/03/11
●A 0635/16/02	●P 1335/19/04	**2009** ☊ 311.0°	
☽P 1130/28/07	☽T 2030/04/05	●A 0800/26/01	**2014** ☊ 214.3°
●T 1104/11/08	●P 0300/14/10	●T 0235/22/07	☽T 0750/15/04
	☽T 0305/28/10	☽P 1925/31/12	●A 0600/29/04
2000 ☊ 125.1°			☽T 1100/08/10
☽T 0445/21/01	**2005** ☊ 28.3°	**2010** ☊ 291.6°	●P 2150/23/10
●P 1250/05/02	●T 2035/08/04	●A 0710/15/01	
●P 1930/01/07	●A 1035/03/10	☽P 1140/26/06	**2015** ☊ 194.9°
☽T 1400/16/07	☽P 1205/17/10	●P 1935/11/07	●T 0950/20/03
●P 0215/31/07		☽T 0815/21/12	☽P 1200/04/04
●P 1735/25/12			●P 0700/13/09
			☽T 0250/28/09

Learning to recognize the constellations

The easiest way to learn your way around the stars is to identify particular constellations and to use those as a key to finding the rest. Such "key" constellations can be found in two directions, looking toward the poles, and looking toward the equator.

The "equatorial" maps in this atlas show the constellations as they appear when close to the meridian, looking north and south. They are drawn to a very small scale and the user should bear this in mind when relating the maps to the very large scale of the sky. One side of a map to the other represents just over 80°, top to bottom represents more than 120°.

Looking with one eye at a hand held at arm's length gives a rough measure of angular size in the sky: a thumbnail covers 2°, a clenched fist spans 9–10°, a handspan (thumb to tip of little finger with the fingers spread) covers 20–24°.

Looking toward the poles

The key northern groups are Ursa Major and Cassiopeia. The main stars of Ursa Major form the shape of a ladle, and in America this group is known as the Big Dipper. In Britain it is known as the Plough. An imaginary line drawn through Merak and Dubhe (β and α UMa) leads to Polaris (α UMi), the Pole Star. Cassiopeia is found on the other side of the pole from Ursa Major, and looks like a giant W or M, depending on the angle.

The region around the south celestial pole has no bright stars in it, so there is no convenient pole star in the southern sky. The best known group in the southern sky is Crux, the Southern Cross. Crux is circumpolar from southern Australia and most of New Zealand. Fifteen degrees of arc – just under a handspan – east of Crux is the bright pair Agena and Rigil Kent (β and α Cen), 5° apart. Crux and these two stars stand 30° to one side of the celestial pole, while Achernar (α Eri) stands 30° to the other side. The solitary bright star Canopus (α Car) also stands a similar distance away from the pole, halfway along the arc separating Achernar from Crux.

Northern hemisphere, *looking south*

• In winter (December onward, Map 2) the key group is Orion. NW of Orion are the Pleiades, a smudgy-looking cluster of stars, and the orange star Aldebaran, the eye of the Bull (Taurus). Sirius, the brightest star in the sky, rises in the SE.

• In spring (March onward, Map 4) Leo (containing the bright bluish star Regulus) is seen high in the southern sky, with Ursa Major "upside down" on the other side of the zenith. Continuing the handle of Ursa Major down in a curve leads to Arcturus, the yellowish star in Boötes.

• In summer (June onward, Map 6) the key star Vega (blue–green) is high in the east. Yellowish Arcturus (map 5) has passed the meridian, and Corona Borealis lies between. The red star Antares (in Scorpius) is low on the southern horizon.

• In late summer (August, Map 6) find the large-scale grouping known as the Summer Triangle: Vega (in Lyra), Deneb (in Cygnus) and Altair (in Aquila).

• In fall (September onward, Map 1) the key group is the Square of Pegasus. The Summer Triangle is still visible.

Southern hemisphere, *looking north*

Hold the maps upside down!

• In summer (December onward, Map 2) the key grouping is Orion, with the bright star Sirius above it to the east. On the other side of the zenith the Large Magellanic Cloud lies between Canopus and the pole.

• In fall (March onward, Map 4) Regulus is in the northern sky after dark. Corvus approaches the zenith at midnight, with blue–green Spica (Virgo) following.

• In winter (June onward, Map 5) the key star is red Antares in Scorpius, with the rich starfields of Sagittarius following.

• In spring (September onward, Map 1) the key star Fomalhaut (α PsA) is close to the zenith. Fomalhaut heads a giant vee: 28° NE is Difda (β Cet), 23° SE is Ankaa (α Phe), and 20° SE of Ankaa is Achernar (α Eri).

21

The constellation names

When the constellation name follows a star's name, letter or number, the genitive form of the name is used, thus: Deneb Cygni, α Cygni, 50 Cygni. But abbreviations of the constellation names (or rather, of their genitive forms) are increasingly used in printed references to stars: for example, α Cyg.

Abbr.	Name	Genitive	English Form
And	Andromeda	Andromedae	Andromeda
Ant	Antlia	Antliae	The Air Pump
Aps	Apus	Apodis	The Bird of Paradise
Aqr	Aquarius	Aquarii	The Water Carrier
Aql	Aquila	Aquilae	The Eagle
Ara	Ara	Arae	The Altar
Ari	Aries	Arietis	The Ram
Aur	Auriga	Aurigae	The Charioteer
Boo	Boötes	Boötis	The Herdsman
Cae	Caelum	Caeli	The Graving Tool
Cam	Camelopardalis	Camelopardalis	The Giraffe
Cnc	Cancer	Cancri	The Crab
CVn	Canes Venatici	Canum Venaticorum	The Hunting Dogs
CMa	Canis Major	Canis Majoris	The Greater Dog
CMi	Canis Minor	Canis Minoris	The Lesser Dog
Cap	Capricornus	Capricorni	The Sea Goat
Car	Carina	Carinae	The Keel
Cas	Cassiopeia	Cassiopeiae	Cassiopeia
Cen	Centaurus	Centauri	The Centaur
Cep	Cepheus	Cephei	Cepheus
Cet	Cetus	Ceti	The Whale
Cha	Chamaeleon	Chamaeleontis	The Chamaeleon
Cir	Circinus	Circini	The Compasses
Col	Columba	Columbae	The Dove
Com	Coma Berenices	Comae Berenicis	Berenice's Hair
CrA	Corona Australis	Coronae Australis	The Southern Crown
CrB	Corona Borealis	Coronae Borealis	The Northern Crown
Crv	Corvus	Corvi	The Crow
Crt	Crater	Crateris	The Cup
Cru	Crux	Crucis	The Southern Cross
Cyg	Cygnus	Cygni	The Swan
Del	Delphinus	Delphini	The Dolphin
Dor	Dorado	Doradus	The Swordfish
Dra	Draco	Draconis	The Dragon
Equ	Equuleus	Equulei	The Little Horse
Eri	Eridanus	Eridani	The River Eridanus
For	Fornax	Fornacis	The Furnace
Gem	Gemini	Geminorum	The Twins
Gru	Grus	Gruis	The Crane
Her	Hercules	Herculis	Hercules

Abbr.	Name	Genitive	English Form
Hor	Horologium	Horologii	The Pendulum Clock
Hya	Hydra	Hydrae	The Water Snake
Hyi	Hydrus	Hydri	The Lesser Water Snake
Ind	Indus	Indi	The Indian
Lac	Lacerta	Lacertae	The Lizard
Leo	Leo	Leonis	The Lion
LMi	Leo Minor	Leonis Minoris	The Smaller Lion
Lep	Lepus	Leporis	The Hare
Lib	Libra	Librae	The Scales
Lup	Lupus	Lupi	The Wolf
Lyn	Lynx	Lyncis	The Lynx
Lyr	Lyra	Lyrae	The Lyre
Men	Mensa	Mensae	The Table Mountain
Mic	Microscopium	Microscopii	The Microscope
Mon	Monoceros	Monocerotis	The Unicorn
Mus	Musca	Muscae	The Fly
Nor	Norma	Normae	The Level
Oct	Octans	Octantis	The Octant
Oph	Ophiuchus	Ophiuchi	The Serpent Holder
Ori	Orion	Orionis	Orion
Pav	Pavo	Pavonis	The Peacock
Peg	Pegasus	Pegasi	Pegasus
Per	Perseus	Persei	Perseus
Phe	Phoenix	Phoenicis	The Phoenix
Pic	Pictor	Pictoris	The Painter
Psc	Pisces	Piscium	The Fishes
PsA	Piscis Austrinus	Piscis Austrini	The Southern Fish
Pup	Puppis	Puppis	The Stern
Pyx	Pyxis	Pyxidis	The Mariner's Compass
Ret	Reticulum	Reticuli	The Net
Sge	Sagitta	Sagittae	The Arrow
Sgr	Sagittarius	Sagittarii	The Archer
Sco	Scorpius	Scorpii	The Scorpion
Scl	Sculptor	Sculptoris	The Sculptor
Sct	Scutum	Scuti	The Shield
Ser	Serpens*	Serpentis	The Serpent
Sex	Sextans	Sextantis	The Sextant
Tau	Taurus	Tauri	The Bull
Tel	Telescopium	Telescopii	The Telescope
Tri	Triangulum	Trianguli	The Triangle
TrA	Triangulum Australe	Trianguli Australis	The Southern Triangle
Tuc	Tucana	Tucanae	The Toucan
UMa	Ursa Major	Ursae Majoris	The Great Bear
UMi	Ursa Minor	Ursae Minoris	The Little Bear
Vel	Vela	Velorum	The Sails
Vir	Virgo	Virginis	The Virgin
Vol	Volans	Volantis	The Flying Fish
Vul	Vulpecula	Vulpeculae	The Little Fox

* The constellation of Serpens has been split into two halves, which are sometimes denoted separately:

| SerCp | Serpens Caput | Serpentis Caput | The Serpent's Head |
| SerCd | Serpens Cauda | Serpentis Cauda | The Serpent's Tail |

390° Celestial Equator Map

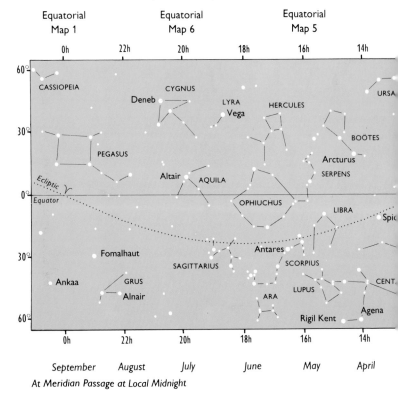

Equatorial Map 1   Equatorial Map 6   Equatorial Map 5

0h   22h   20h   18h   16h   14h

60°

CASSIOPEIA

Deneb   CYGNUS   LYRA   HERCULES   URSA
                    Vega

30°   BOÖTES

PEGASUS   Arcturus
                          SERPENS

Ecliptic

0°   Altair   AQUILA
Equator   OPHIUCHUS   LIBRA   Spic

Fomalhaut   Antares

30°   SCORPIUS
        SAGITTARIUS   CENT

Ankaa   GRUS   LUPUS
        Alnair   ARA   Agena

60°   Rigil Kent

0h   22h   20h   18h   16h   14h

September   August   July   June   May   April

At Meridian Passage at Local Midnight

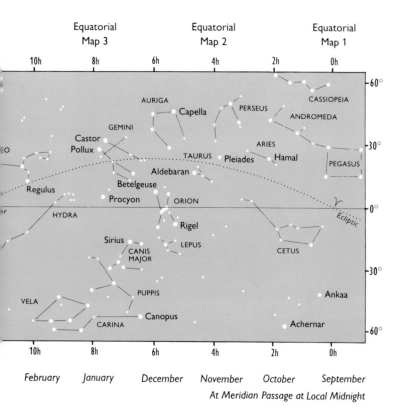

Equatorial Map 3   Equatorial Map 2   Equatorial Map 1

February   January   December   November   October   September

At Meridian Passage at Local Midnight

North Circle 100°

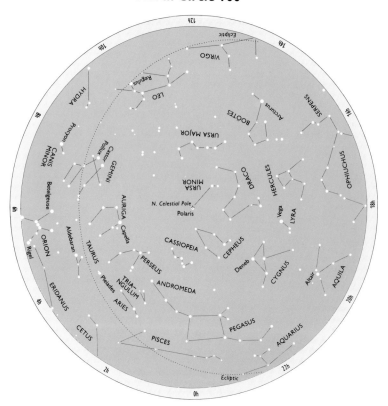

Greek Alphabet			
α alpha	η eta	ν nu	τ tau
β beta	θ theta	ξ xi	υ upsilon
γ gamma	ι iota	ο omicron	φ phi
δ delta	κ kappa	π pi	χ chi
ε epsilon	λ lambda	ϱ rho	ψ psi
ζ zeta	μ mu	σ sigma	ω omega

North Polar Map

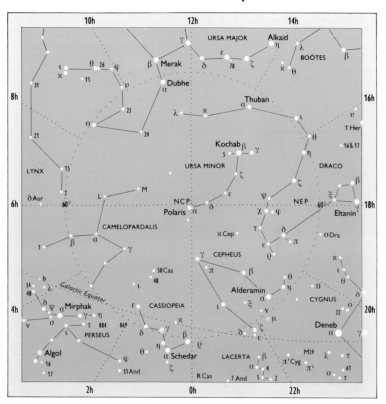

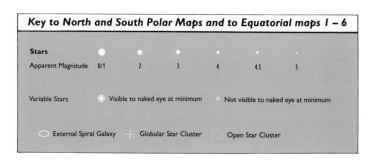

Equatorial Map 6

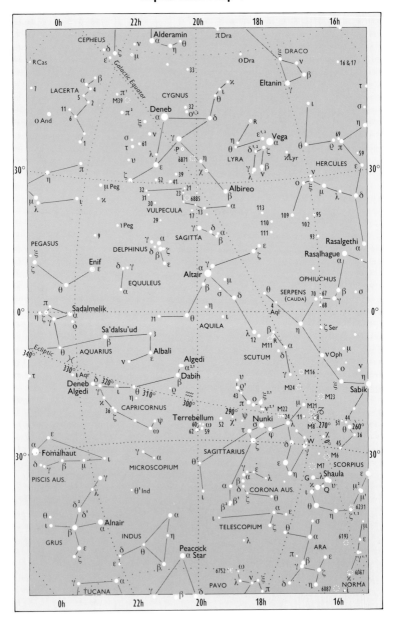

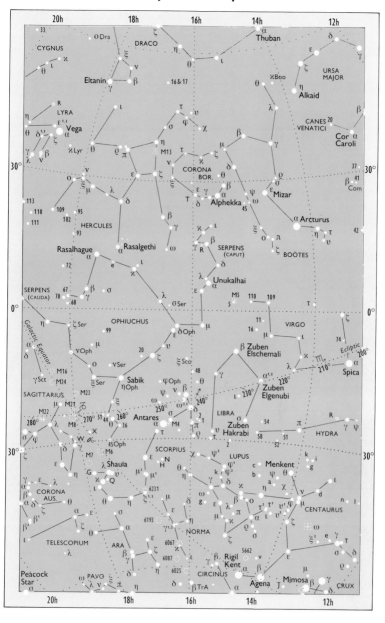

Equatorial Map 4

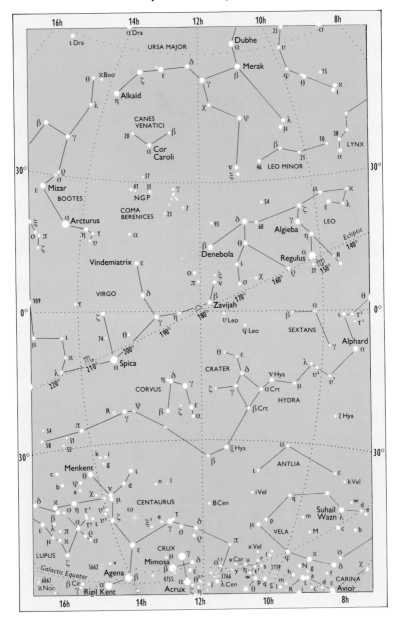

Equatorial Map 3

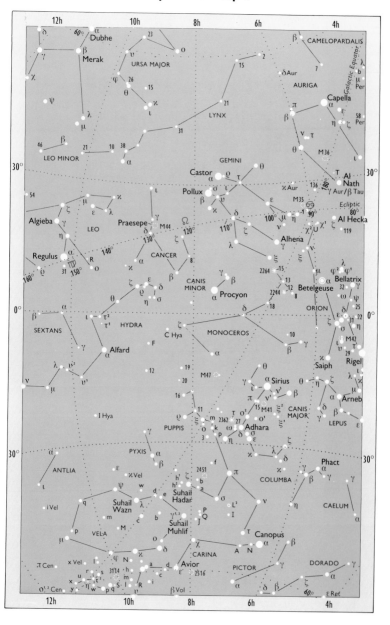

Equatorial Map 2

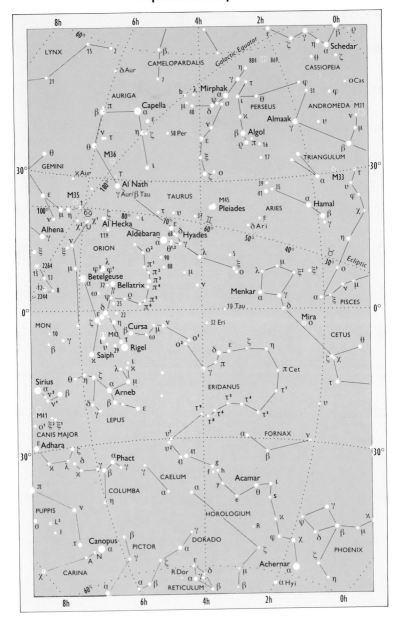

Equatorial Map 1

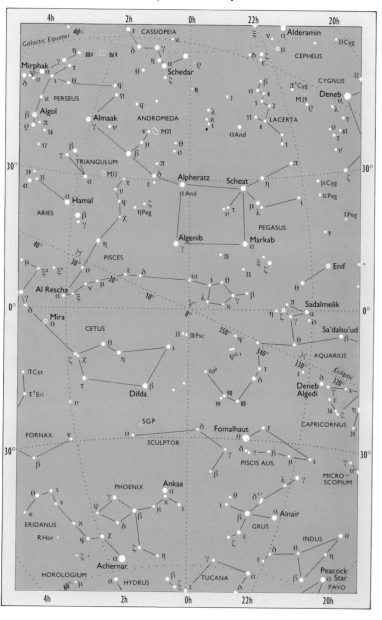

South Polar Map

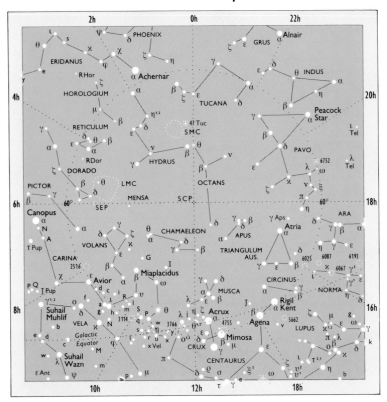

Key to North and South Polar Maps and to Equatorial maps 1 – 6

Stars

Apparent Magnitude   0/1   2   3   4   4.5   5

Variable Stars   ◉ Visible to naked eye at minimum   ○ Not visible to naked eye at minimum

◯ External Spiral Galaxy   ✛ Globular Star Cluster   ⊙ Open Star Cluster

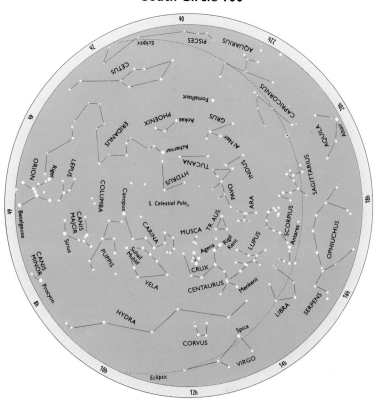

Greek Alphabet			
α alpha	η eta	ν nu	τ tau
β beta	θ theta	ξ xi	υ upsilon
γ gamma	ι iota	ο omicron	φ phi
δ delta	κ kappa	π pi	χ chi
ε epsilon	λ lambda	ϱ rho	ψ psi
ζ zeta	μ mu	σ sigma	ω omega

Variable stars

Variable stars are stars whose brightness (listed as apparent magnitude) varies over time. This variation may take place in a regular, semiregular, irregular, or completely unpredictable way. Intrinsic variables (the supreme example is Mira, o Cet) are variable because of instability or cyclical fuel economies. Some intrinsic variables are of one spectral type/luminosity class at maximum and of another at minimum. In the table both are shown, separated by a dash. Extrinsic (eclipsing) variables (such as Algol, β Per) are multiple systems in which one member periodically eclipses another. In the table the period of variation is shown in days (average periods in italics).

Name	Max.	Min.	Period	Spectrum	Variability type
T Aur	4.1	15.5	−	pec	nova + eclipsing
η Car	−0.8	7.9	−	pec	eruptive (S Dor type)
R Car	3.9	10.5	308.7	M4e-M8e	pulsating (Mira type)
S Car	4.5	9.9	149.5	K5e-M6e	pulsating (Mira type)
Cas	4.1	6.2	320	F8Ia-K0Ia-o	semiregular pulsating
γ Cas	1.6	3.0	−	B0.5IVpe	irregular eruptive
R Cas	4.7	13.5	430.5	M6e-M10e	pulsating (Mira type)
μ Cep	3.43	5.1	730	M2eIa	semiregular pulsating
o Cet	2.0	10.0	331.96	M5e-M9e	pulsating (Mira)
T CrB	2.0	10.8	−	M3III+pec(Nova)	recurrent nova
χ Cyg	3.3	14.2	408.05	S6, 2-S10,4e	pulsating (Mira type)
P Cyg	3	6	−	B1Iapeq	eruptive (S Dor type)
R Dor	4.8	6.6	388	M8IIIe	semiregular pulsating
α Her	2.74	4.0	−	M5Ib-II	semiregular pulsating
R Hor	4.7	14.3	407.6	M5e-M8eII-III	pulsating (Mira type)
R Hya	3.5	10.9	388.87	M6e-M9es	pulsating (Mira type)
R Leo	4.4	11.3	309.95	M6e-M9.5IIIe	pulsating (Mira type)
β Lyr	3.25	4.36	12.91	B8II-IIIep	eclipsing
R Lyr	3.88	5.0	46	M5III	semiregular pulsating
α Ori	0.0	1.3	2335	M1-M2Ia-Ibe	semiregular puls.
U Ori	4.8	13.0	368.3	M6e-M9.5e	pulsating (Mira type)
ε Peg	0.7	3.5	−	K2Ib	slow irregular pulsating
β Per	2.12	3.39	2.87	B8V	eclipsing (Algol)
L² Pup	2.6	6.2	140.6	M5IIIe-M6IIIe	semiregular pulsating
R Sct	4.2	8.6	146.5	G0Iae-K2pIbe	pulsating (RV Tau type)
R Ser	5.16	14.4	56.41	M5IIIe-M9e	pulsating (Mira type)

The brightest stars

Spectrum: A double spectrum (for example, G2V+K1V) denotes members of a binary system that appears single to the naked eye (list shows combined magnitude); M_v: values not certain; **Dist.:** distances are theoretical beyond 50 light-years; m: multiple system; (n): number in system; v: variable.

Star		m_v	Spectrum	M_v	Dist. ly	Notes
Sirius	α CMa	−1.46	A1Vm	1.4	8.8	m(4)
Canopus	α Car	−0.72	F0Ib	−4.7	200	
Rigil Kent	α¹·²Cen	−0.27	G2V+K1V	−	4.25	m(3)
Arcturus	α Boo	−0.04	K2IIIp	−0.3	36	−
Vega	α Lyr	0.03	A0V	0.5	26	m(4)
Capella	α Aur	0.08	G5IIIe+G0III	−	46	m(10)
Rigel	β Ori	0.12	B8Ia	−7.4	1050	m(4)
Procyon	α CMi	0.34	F5IV–V	2.6	11.5	m(5)
Achernar	α Eri	0.46	B3Vpe	−1.7	90	
Betelgeuse	α Ori	0.50	M1-2Ia-Ibe	−	420	m(6),v
Agena	β Cen	0.60	B1III	−4.7	360	m(2)
Altair	α Aql	0.77	A7IV–V	−	17	m(3)
Aldebaran	α Tau	0.85	K5III	−0.7	70	m(6)
Acrux	α¹·² Cru	0.87	B0.5IV+B1V	−	520	m(3)
Antares	α Sco	0.96	M1.5Ia-Ib+B4Ve	−	600	m(2),v
Spica	α Vir	0.98	B1III-IV+B2V	−	520	m(5),v
Pollux	β Gem	1.14	K0III	1.0	6	m(7)
Fomalhaut	α PsA	1.16	A3V	2.0	22	
Mimosa	β Cru	1.20	B0.5III	−5.0	490	m(3),v
Deneb	α Cyg	1.25	A2Ia	−7.5	1650	m(2)
Regulus	α Leo	1.35	B7V	−0.9	90	m(4)
Adhara	ε CMa	1.50	B2II	−4.4	490	m(2)
Castor	α Gem	1.58	A1V+Am	−	50	m(4)
Gacrux	γ Cru	1.63	M3II	−0.9	105	m(3)
Shaula	λ Sco	1.63	B2IV+B	−	180	m(3)
Bellatrix	γ Ori	1.64	B2III	−3.6	360	m(2)
Al Nath	β Tau	1.65	B7III	−1.6	150	m(2)
Alnilam	ε Ori	1.70	B0Iae	−6.2	1200	m(2)
Miaplacidus	β Car	1.70	A2IV	−0.6	55	
Al Nath	α Gru	1.74	B5IV	−0.2	65	m(2)
Alnitak	ζ Ori	1.75	B0III+O9.5Ib	−	1000	m(4)
Alioth	ε UMa	1.77	A0p	−	−	v
Suhail Muhlif	γ Vel	1.78	WR8+O7.5e	−	1500	m(5)

37

Star clusters and galaxies shown on the maps

Mag.: Integrated apparent magnitude for the whole object

Dia.: Approximate diameter of object in arcseconds (3600 arcsec = 1°)

Globular clusters

Designation	Name	RA		Decl.		Const.	Dia.	Mag.
		h	min	deg	min		arcsec	
NGC 104	47 Tuc	00	24.1	−72	05	Tuc	30.9	4.03
Visible with naked eye								
NGC 5139	ω Cen	13	26.8	−47	29	Cen	36.3	3.65
Visible with naked eye								
NGC 5904	M5	15	18.6	+02	05	SerCp	17.4	5.75
NGC 6121	M4	16	23.6	−26	32	Sco	26.3	5.93
NGC 6205	M13	16	41.7	+36	28	Her	16.6	5.86
NGC 6656	M22	18	36.4	−23	54	Sgr	24.0	5.10
NGC 6752	Dun 295	19	10.9	−59	59	Pav	20.4	5.4

Open clusters

Designation	Name	RA		Decl.		Const.	Dia.	Mag.
		h	min	deg	min		arcsec	
NGC 869	h Per	02	19.0	+57	09	Per	30	4.3
NGC 884	c Per	02	22.4	+57	07	Per	30	4.4
NGC 869 & 884 make up the Double Cluster, naked-eye object								
Pleiades	M45	03	47	+24	07	Tau	110	1.2
The Seven Sisters naked-eye cluster 390 ly away								
Hyades		04	25	+16		Tau	660	0.5
Naked-eye, nearest cluster (150 ly), Aldebaran not member								
NGC 1976	M42	05	35.4	−05	27	Ori	66	4
Trapezium cluster and Great Orion Nebula, naked eye object								
NGC 1960	M36	05	36.1	+34	08	Aur	12	6.0
Mid placed of three clusters, all binocular objects								
NGC 2168	M35	06	08.9	+24	20	Gem	28	5.1
NGC 2244		06	32.4	+04	52	Mon	24	4.8
Associated with the Rosette Nebula								
NGC 2264		06	41.1	+09	53	Mon	20	3.9
Associated with Cone Nebula (nebula visible only in photographs)								
NGC 2287	M41	06	47.0	−20	44	CMa	38	4.5
NGC 2362		07	20.1	−13	08	CMa	8	4.1
NGC 3422	M47	07	36.6	−14	30	Pup	30	4.4
NGC 2451		07	45.4	−37	58	Pup	45	3.6

Designation	Name	RA h min	Decl. deg min	Const.	Dia. arcsec	Mag.
NGC 2516		07 58.3	−60 52	Car	30	3.8
NGC 2632	M44	08 40.1	+19 59	Cnc	95	3.1
Praesepe or Beehive Cluster; naked-eye						
NGC 3114		10 02.7	−60 07	Car	35	4.2
IC 2602 q Car cluster		10 43.2	−64 24	Car	50	2.77
Naked-eye						
NGC 3766		11 36.1	−61 37	Cen	12	5.3
NGC 4755 k Cru cluster		12 53.6	−60 20	Cru	10	4.2
The famous Jewel-Box Cluster, a binocular object						
NGC 5662		14 35.2	−56 33	Cen	12	5.5
NGC 6025		16 03.7	−60 30	TrA	12	5.1
NGC 6067		16 13.2	−54 13	Nor	13	5.6
NGC 6087		16 18.9	−57 54	Nor	12	5.4
NGC 6193		16 41.3	−48 46	Ara	15	5.2
NGC 6231		16 54	−41 48	Sco	15	2.6
NGC 6405	M6	17 40.1	−32 13	Sco	15	4.2
NGC 6475	M7	17 53.9	−34 49	Sco	80	3.3
NGC 6494	M23	17 56.8	−19 01	Sgr	27	5.5
NGC 6531	M21	18 04.6	−22 30	Sgr	13	5.9
NGC 6530	M8	18 04.8	−24 20	Sgr	15	4.6
Naked-eye cluster associated with the Lagoon Nebula						
NGC 6611	M16	18 18.8	−13 47	SerCd	7	6.0
NGC 6603	M24	18 15	−18 30	Sgr		
Contains a true cluster, but essentially a rich star field						
NGC 6618	M17	18 20.8	−16 11	Sgr	11	6.0
Omega Nebula and associated cluster						
NGC 6705	M11	18 51.1	−06 16	Sct	14	5.8
Wild Duck Cluster, good binocular object						
NGC 6871		20 05.9	+35 47	Cyg	20	5.2
NGC 6885		20 12.0	+26 29	Vul	7	5.7
NGC 7092	M39	21 32.2	+48 26	Cyg	32	4.6

Galaxies

Designation	Name	RA h min	Decl. deg min	Const.	Dia. arcsec	Mag.
NGC 224	M31	00 42.7	+41 16	And	178	3.5
Andromeda Galaxy; spiral galaxy at 2.15 Mly; naked-eye						
NGC 598	M33	01 33.9	+30 39	Tri	62	5.7
Triangulum Galaxy; spiral galaxy at 2.35 Mly; difficult naked-eye object						
Small Magellanic Cloud (SMC)		01	−72	Tuc		
Dwarf irregular galaxy, satellite system to Milky Way at 200,000 ly						
Large Magellanic Cloud (LMC)		05 30	−69	Dor/Men		
As SMC, but at 160,000 ly						

The differences between stars and planets

Stars are very large bodies of gas that create their own light from nuclear fusion processes in their interiors. The Sun is a star. Except for the Sun, all stars are very distant objects. The planets are (comparatively) small objects in orbit round the Sun, and the Earth is one of them. To the naked eye the planets Mercury, Venus, Mars, Jupiter, and Saturn look like very bright stars, but their light is borrowed, reflected from the Sun. The main difference between stars and planets that can be seen by the naked eye is that the stars hold fixed positions relative to each other but planets move: planets get their name from the Greek *astéres planêtai*, "wandering stars."

Inner and outer planets

Mercury and Venus are called the inner planets because their orbits are inside that of the Earth. Mars, Jupiter, and Saturn are outer planets because their orbits lie outside that of the Earth. These are the five planets known to the ancient world, although some early observers did not realize that the morning and evening apparitions of Venus and Mercury were of the same planets, and the phrase "the seven stars" may derive from this.

In 1781 a sixth planet was discovered and called Uranus; this was followed by Neptune in 1846 and by Pluto in 1930. Under suitable conditions an experienced observer can see Uranus with the naked eye; it looks like a very faint star. Uranus is more easily observed with the help of binoculars, detailed tables, and a star map that is very much more detailed than the ones in this book. Neptune and Pluto are telescope objects.

The Earth and all the other planets except Pluto orbit the Sun in much the same plane, and seen from Earth they keep close to the ecliptic. The outer planets, Mars, Jupiter, and Saturn, move in apparent independence of the Sun in an overall easterly direction. The inner planets Mercury and Venus appear tied to the Sun, and are seen only in the western sky after sunset or in the eastern sky before dawn.

The Solar System

m_v	Average brightness at elongation, ($\mathrm{\breve{\varphi}}$,$\mathrm{\varphi}$) or at opposition ($\mathrm{\sigma}$,$\mathrm{2\!\!\!4}$,$\mathrm{\hbar}$,$\mathrm{\mathbb{H}}$,$\mathrm{\Psi}$,P,$\mathrm{\mathbb{D}}$,$\mathrm{\odot}$).
D	Mean distance from Sun, in millions of miles
e	Eccentricity of elliptical orbit
i_{orb}	Inclination of orbit relative to ecliptic
D_{pol}	Polar diameter in miles
D_{equ}	Equatorial diameter in miles
i_{equ}	Inclination of equator to plane of orbit
M_{terr}	Mass expressed in Earth Masses
ϱ	Density, expressed in grams per cubic centimeter
P	Sidereal period – time to complete one orbit relative to fixed stars
S	Synodic period – time to return to same position relative to the Sun as observed from Earth
moons	Number of satellite moons in orbit round planet

	m_v	D	e	i_{orb}	D_{pol}	D_{equ}
$\breve{\varphi}$ Mercury	−0.2	36.0	0.2056	7.004°	–	3030
φ Venus	−4.2	67.2	0.0068	3.394°	–	7523
$\oplus$ Earth	–	93.0	0.0167	–	7900	7926
σ Mars	−2.0	141.6	0.0934	1.850°	–	4220
$2\!\!\!4$ Jupiter	−2.5	483.7	0.0485	1.306°	83,100	89,400
$\hbar$ Saturn	0.7	886.7	0.0556	2.490°	67,600	74,900
$\mathbb{H}$ Uranus	5.5	1783.1	0.0472	0.773°	–	31,800
Ψ Neptune	7.9	2794.2	0.0086	1.775°	–	31,400
P Pluto	14.9	3665.5	0.2502	17.143°	–	1400
$\mathbb{D}$ Moon	−12.5	Diameter 2160 miles		Distance from Earth 238,850 miles		
$\odot$ Sun	−27	Diameter 865,000 miles				

	i_{equ}	M_{terr}	ϱ	P	S	moons
Mercury	0.0°	0.056	5.6	87.97d	115.88d	0
Venus	177.3°	0.8148	5.1	224.70d	583.92d	0
Earth	23.45°	1	5.52	365.256d	–	1
Mars	23.19°	0.1078	3.97	686.98d	779.94d	2
Jupiter	3.13°	317.82	1.30	11.86y	398.88d	16
Saturn	26.72°	95.1	0.68	29.46y	378.09d	17
Uranus	97.86°	14.5	1.3	84.01y	369.66d	14
Neptune	29.56°	17.2	1.8	164.79y	367.49d	2
Pluto	117.56°	0.925	2	248.5y	366.73d	1
Moon	–	0.0123	3.3	–	–	–
Sun	–	332,946	1.409	–	–	–

Direct and retrograde motion

Observed from the Earth the planets show an overall motion from west to east along the path of the ecliptic. This easterly movement is called direct motion, and it reflects the true direction of the planets' orbits round the Sun. However, the closer a planet is to the Sun the faster it orbits, and as the Earth passes between the Sun and an outer planet, overtaking it on the inside, the planet goes through a period when it appears to move backward against the fixed stars – east to west – in apparent, retrograde, motion. Retrograde motion is also seen when an inner planet passes between the Earth and the Sun, but because the inner planets are mostly observed in twilight, retrogression against the fixed stars is not very obvious.

Behavior and appearance of the inner planets

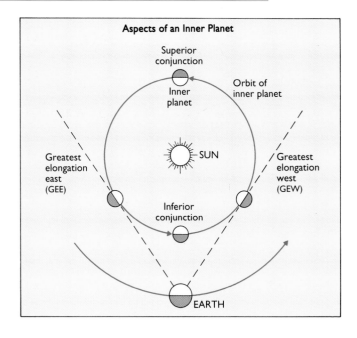

Aspects of an Inner Planet

At superior conjunction an inner planet lies on the far side of the Sun and cannot be seen. As it moves into elongation east ("elongation" means angular separation from the Sun) it becomes visible as an evening object, one that sets in the west after the Sun. After greatest elongation east (GEE) it moves back toward the Sun. At inferior conjunction it passes between the Earth and the Sun and cannot be seen. As it moves into elongation west it reappears as a morning object, rising in the east before the Sun. After GEW it turns back toward the Sun and disappears on its way toward superior conjunction.

Apparitions of Mercury

The greatest elongation of Mercury ranges from 18° to 28°, making some apparitions more difficult than others. Searches should be made for about 30 minutes, starting 40 minutes after sunset (evening apparition, easiest in spring) or 70 minutes before sunrise (morning apparition, easiest in fall). The first search should be made a few days before greatest elongation. The sky needs to be clear almost to the level of the horizon. Mercury appears as a starlike object 6–18° above the horizon in the region blanched by the hidden Sun.

Mercury: greatest elongation east (evening apparition)							
1996	Jan 3	2001	Jan 29	2006	Feb 23	2011	Mar 22
	Apr 22		May 21		Jun 20		Jul 19
	Aug 21		Sep 18		Oct 17		Nov 13
	Dec 15	2002	Jan 12	2007	Feb 7	2012	Mar 4
1997	Apr 5		May 4		Jun 2		Jun 30
	Aug 3		Aug 31		Sep 28		Oct 26
	Nov 28		Dec 26	2008	Jan 21	2013	Feb 16
1998	Mar 19	2003	Apr 16		May 13		Jun 11
	Jul 16		Aug 14		Sep 10		Oct 9
	Nov 11		Dec 9	2009	Jan 4	2014	Jan 31
1999	Mar 3	2004	Mar 29		Apr 25		May 24
	Jun 28		Jul 26		Aug 23		Sep 21
	Oct 24		Nov 20		Dec 19	2015	Jan 14
2000	Feb 14	2005	Mar 12	2010	Apr 8		May 6
	Jun 8		Jul 8		Aug 6		Sep 3
	Oct 6		Nov 3		Dec 1		Dec 28

Mercury: greatest elongation west (morning apparition)								
1996	Feb 11	**2001**	Mar 10	**2006**	Apr 8	**2011**	Jan 8
	Jun 10		Jul 9		Aug 6		May 7
	Oct 3		Oct 29		Nov 24		Sep 2
1997	Jan 23	**2002**	Feb 20	**2007**	Mar 21		Dec 22
	May 22		Jun 21		Jul 20	**2012**	Apr 17
	Sep 16		Oct 12		Nov 8		Aug 16
1998	Jan 6	**2003**	Feb 3	**2008**	Mar 3		Dec 4
	May 4		Jun 3		Jul 1	**2013**	Mar 31
	Aug 31		Sep 26		Oct 21		Jul 30
	Dec 19	**2004**	Jan 16	**2009**	Feb 12		Nov 17
1999	Apr 16		May 14		Jun 13	**2014**	Mar 13
	Aug 14		Sep 9		Oct 5		Jul 12
	Dec 3		Dec 29	**2010**	Jan 26		Nov 1
2000	Mar 28	**2005**	Apr 26		May 25	**2015**	Feb 24
	Jul 27		Aug 23		Sep 19		Jun 24
	Nov 14		Dec 12				Oct 15

Mercury at elongation varies in brightness from $m_v +1$ to -1, on average rather brighter than any star that might appear nearby. In an evening observation Mercury is likely to be the first starlike object to become visible close to the western horizon, and in a morning observation the last starlike object to rise in the lightening eastern sky.

From one evening apparition of Mercury to the next takes four months (synodic period 115.88 days) From GEE to GEW takes 45 days, while GEW to GEE takes 75 days.

Apparitions of Venus

From one evening apparition of Venus to the next takes 19 months (synodic period 583.92 days). GEE to GEW takes 20 weeks, and GEW to GEE takes 63 weeks.

Venus becomes visible as an evening object four months before GEE and becomes easy to find two months before. At GEE it reaches 45–47° of separation from the Sun, and brightens. To the naked eye it looks like a brilliant white starlike object, but through binoculars the planet is seen in a crescent phase. After GEE the planet turns back toward the Sun, but continues to brighten for five weeks, reaching a maximum of m_V -4.4 (15 times brighter than Sirius) before accelerating in apparent motion toward the Sun, and fading abruptly from sight.

Three weeks after inferior conjunction the planet appears in the morning sky and with equal abruptness flares up in two weeks to m_V -4.4, reached five weeks before GEW. After GEW the planet slowly turns back toward the Sun, fades, and disappears after about four months.

Venus: greatest elongations east (evening) and west (morning)					
GEE		**GEW**	**GEE**		**GEW**
Mar 20	**1996**	Aug 20	–	**2006**	Mar 26
Nov 3	**1997**	–	Jun 9	**2007**	Oct 27
–	**1998**	Mar 28	–	**2008**	–
Jun 11	**1999**	Oct 30	Jan 14	**2009**	Jun 6
–	**2000**	–	Aug 19	**2010**	–
Jan 17	**2001**	Jun 8	–	**2011**	Jan 8
Aug 20	**2002**	–	Mar 25	**2012**	Aug 16
–	**2003**	Jan 8	Oct 30	**2013**	–
Mar 29	**2004**	Aug 21	–	**2014**	Mar 22
Nov 3	**2005**	–	Jun 4	**2015**	Oct 25

Behavior and appearance of the outer planets

At conjunction the outer planets, Mars, Jupiter, and Saturn, are hidden by the Sun. They are first visible a month (two months for Mars) after conjunction as morning objects that rise in the east before dawn. On each morning that follows the planet rises earlier. At quadrature west the planet rises at midnight and reaches the meridian at dawn. At opposition it rises at sunset, passes the meridian at midnight, and sets at dawn.

When the planet reaches quadrature east it comes out on the meridian at sunset, and sets at midnight. After quadrature east the planet appears progressively lower in the western sky after sunset, until a month or so before conjunction, when it becomes lost in the afterglow of the setting Sun.

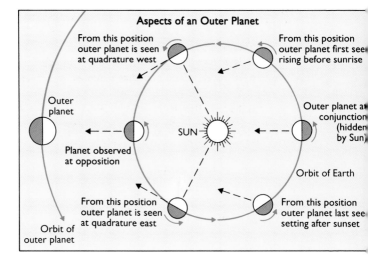

Aspects of an Outer Planet

From this position outer planet is seen at quadrature west

From this position outer planet first seen rising before sunrise

Outer planet

Outer planet at conjunction (hidden by Sun)

SUN

Planet observed at opposition

Orbit of Earth

From this position outer planet is seen at quadrature east

From this position outer planet last seen setting after sunset

Orbit of outer planet

The outer planets Mars, Jupiter, and Saturn are at their brightest at opposition, when the Earth is nearest to them. They are at their most conveniently placed for evening observation from this time through to quadrature east. Over this period they are seen rising in the east in the evening.

For most of each apparition the planet shows an eastward (direct) motion against the stars but around opposition it retrogrades (see page 42). The retrograde motion begins at a point somewhere between quadrature west and opposition, and returns to direct motion at a point between opposition and quadrature east.

Jupiter and Mars at opposition are easily distinguished from stars: Jupiter at opposition (m_V -2.3 to -2.9) is brighter than any star, and Mars (m_V -1.0 to -2.8) is far brighter than any star in whose company it might appear. Jupiter is cream-white, Mars is conspicuously red. Saturn appears no brighter than a very bright star ($m_V = 0.7$) but it has a distinctly yellow color. Saturn moves so little from one year to the next that the observer has only to identify the planet once in order to find it again the next year.

Oppositions of Mars

Mars returns to opposition in about 780 days (2 years, 7 weeks). The planet's orbit is markedly eccentric, so that oppositions range from 40° to 70° east of the one that came before. The planet completes a full circuit of the ecliptic between oppositions (so the distance it travels is in fact 400° to 430°). Retrograde motion starts five weeks before opposition, lasts 10 weeks, and covers 15°.

Date	Ecliptic	Date	Ecliptic	Date	Ecliptic
Mar 17 **1997**	176°	Aug 28 **2003**	335°	Jan 29 **2010**	130°
Apr 24 **1999**	214°	Nov 7 **2005**	45°	Mar 3 **2012**	154°
Jun 13 **2001**	263°	Dec 24 **2007**	93°	Apr 8 **2014**	199°

Oppositions of Júpiter

Jupiter returns to opposition one month later each year, with each opposition 30° east of the one that preceded it (on average both the period and the motion are slightly greater than these values). Retrograde motion starts eight weeks before opposition, lasts 16 weeks, and covers 10°.

Date	Ecliptic	Date	Ecliptic	Date	Ecliptic
Jul 4 **1996**	284°	Feb 2 **2003**	130°	Sep 21 **2010**	258°
Aug 9 **1997**	317°	Mar 4 **2004**	164°	Oct 29 **2011**	35°
Sep 16 **1998**	353°	Apr 3 **2005**	194°	Dec 3 **2012**	71°
Oct 23 **1999**	30°	May 4 **2006**	224°	– **2013**	–
Nov 28 **2000**	66°	Jun 5 **2007**	255°	Jan 5 **2014**	105°
– **2001**	–	Jul 9 **2008**	287°	Feb 6 **2015**	138°
Jan 1 **2002**	101°	Aug 14 **2009**	322°		

Oppositions of Saturn

Saturn returns to opposition two weeks later each year, with each opposition found approximately 13° east of the one that came before. Retrograde motion starts 10 weeks before opposition, lasts 20 weeks, and covers 7°.

Date	Ecliptic	Date	Ecliptic	Date	Ecliptic
Sep 26 **1996**	4°	Dec 31 **2003**	100°	Mar 22 **2010**	181°
Oct 10 **1997**	17°	– **2004**	–	Apr 3 **2011**	194°
Oct 23 **1998**	30°	Jan 13 **2005**	114°	Apr 15 **2012**	204°
Nov 6 **1999**	44°	Jan 27 **2006**	128°	Apr 28 **2013**	218°
Nov 19 **2000**	57°	Feb 10 **2007**	142°	May 10 **2014**	230°
Dec 4 **2001**	71°	Feb 24 **2008**	155°	May 23 **2015**	242°
Dec 17 **2002**	86°	Mar 8 **2009**	168°		

Meteor showers

Meteors, often called "shooting stars," are visible as sudden streaks of light across the sky. They are produced when a piece of interplanetary material is caught up and vaporized in the atmosphere of the Earth. On rare occasions the fragment is sufficiently large to survive its fall and may be recovered as a meteorite. The Earth's orbit passes through patches of interplanetary dust that produce predictable showers. Each shower appears to radiate from a particular point in the sky. The shower is named after the constellation in which the radiant is found.

Meteors are not to be confused with comets, most of which appear as indistinct objects that take weeks to cross the sky. Points of light seen in steady motion close to the zenith toward either end of the night are produced by sunlight reflected from artificial satellites.

January 3	Quadrantids*	October 20	Orionids
(radiant: RA 15h 30m Decl. 50°N)		November 5	Taurids
April 21	Lyrids	November 16	Leonids
May 4	Aquarids	December 13	Geminids
August 4	Aquarids	December 22	Ursids
August 11	Perseids	(radiant: RA 14h 30m Decl. 80°N)	

*Quadrans was a minor constellation, now abandoned.